Animal Faces
On Snow and Ice

Hannah Kate Sackett

Illustrated by Martin Camm

 Children's Publishing

Columbus, Ohio

 Children's Publishing

This edition published in the United States in 2003 by
McGraw-Hill Children's Publishing,
A Division of The McGraw-Hill Companies
8787 Orion Place
Columbus, Ohio 43240-4027

www.MHkids.com

Library of Congress Cataloging-in-Publication Data is on file with the publisher.

Created and produced by Firecrest Books Ltd
in association with Martin Camm and Hannah Sackett.

Copyright © 2002 Firecrest Books Ltd, Martin Camm, and Hannah Sackett.

All rights reserved. No part of this publication may be reproduced, stored in or introduced into a retrieval system, or transmitted in any form or by any means (electronic, mechanical, photocopying, recording or otherwise), without the prior permission of both the copyright owners and the above publisher of this book.

Art and Editorial Direction by Peter Sackett
Edited by Norman Barrett
Edited in the U.S. by Joanna Schmalz and Catherine Stewart
U.S. Production by Tracy Paulus and Nathan Hemmelgarn
Designed by Phil Jacobs
Color Separation by SC International Pte Ltd, Singapore

Printed in Dubai.

ISBN 1-57768-418-4

1 2 3 4 5 6 7 8 9 10 FBL 06 05 04 03 02

Contents

Musk Ox 4
Reindeer 6
Mosquito 8
Snowy Owl 10
Arctic Fox 12
Narwhal 14
Walrus 16
Polar Bear 18
Killer Whale 20
Ice Fish 22
Gray-headed Albatross 24
Elephant Seal 26
Emperor Penguin 28
Facts Behind the Faces 30
Index 33

Musk Ox

In the far north of the world lie the dry, icy lands that fringe the Arctic Ocean. Only the toughest animals can survive there, and among these is the musk ox. This woolly-coated creature has a large head and long, curved horns. Musk oxen use their horns for breaking though the winter ice to reach the grassy ground beneath. They also use their horns as weapons for defending the herd against attacks from dogs and wolves.

Reindeer

Reindeer also make their home in the Arctic. Unlike other deer, both male and female reindeer have antlers, which are large, branched horns. The antlers are most often used by the male reindeer when fighting over a female during the breeding season. Reindeer shed their antlers every winter and grow new ones every spring.

Mosquito

The animals of the Arctic region have to live through long, harsh winters. Spring and summer bring warmth and more food, but they also bring swarms of mosquitoes. These flying insects hatch from eggs laid the year before. The mosquito has a funnel-like mouth, which it uses to suck juices from plants. The female also sucks blood from animals, leaving behind an itchy and irritating bite-mark.

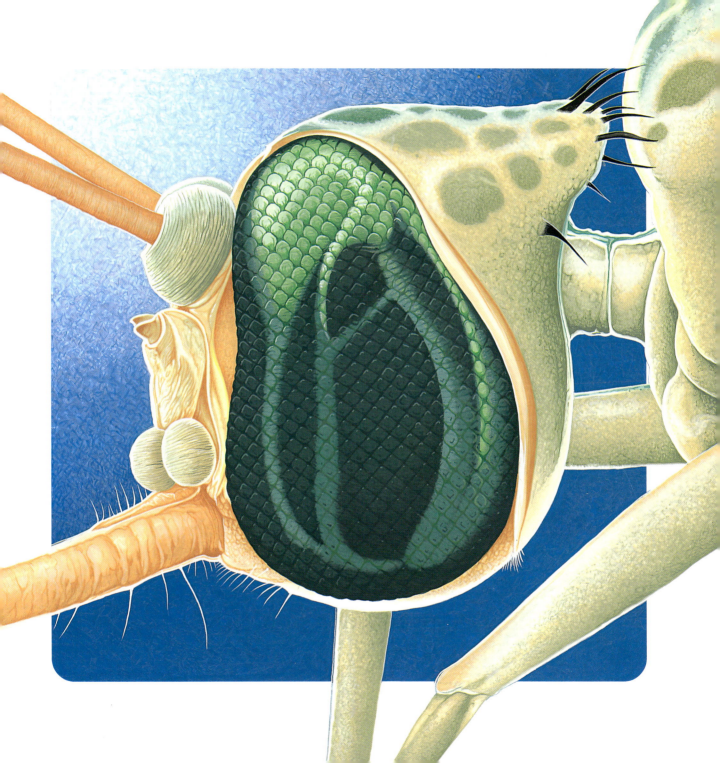

Snowy Owl

The Arctic skies are also home to many birds. Among them is the snowy owl. Unlike other birds, the owl has eyes on the front of its face, rather than on the sides of its head. Its eyeballs are tube-shaped and do not swivel. So, to look around, the owl must move its head. Snowy owls take their name from the color of their feathers, which are gray and white or all white. Their coloring allows them to hunt their prey unseen against the snowy background.

Arctic Fox

The fox is naturally brown in the summer, but as winter approaches, its hair grows in gray and white (right). By the time the snow has settled, the fox is completely white, blending into the wintery backdrop (below). Its ears are small and furry, so the fox does not lose much heat from its head. These features enable the Arctic fox to survive on even the coldest islands of the Arctic.

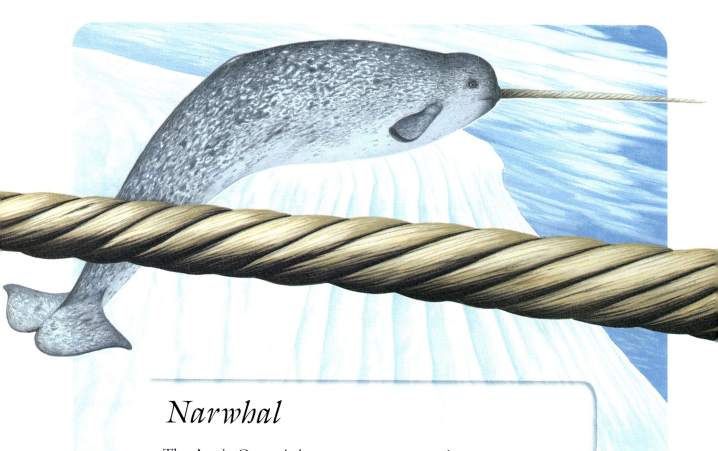

Narwhal

The Arctic Ocean is home to many unusual creatures, including the narwhal. This small whale is known for the long spiral tusk that sticks out from its face. The tusk is formed from a tooth, but no one is sure what it is used for. Not all female narwhals have tusks, but all males have at least one. Some narwhals even have two.

Walrus

The walrus grows two long, impressive tusks, which are actually enlarged canine teeth. It uses them for making air holes in the ice, as well as for rooting out food. The stiff bristles, or whiskers, on its face help the walrus to find clams, crabs, and urchins on the dark seabed. The walrus is at home in the water, but spends most of its time in the shallows, or resting on floating ice.

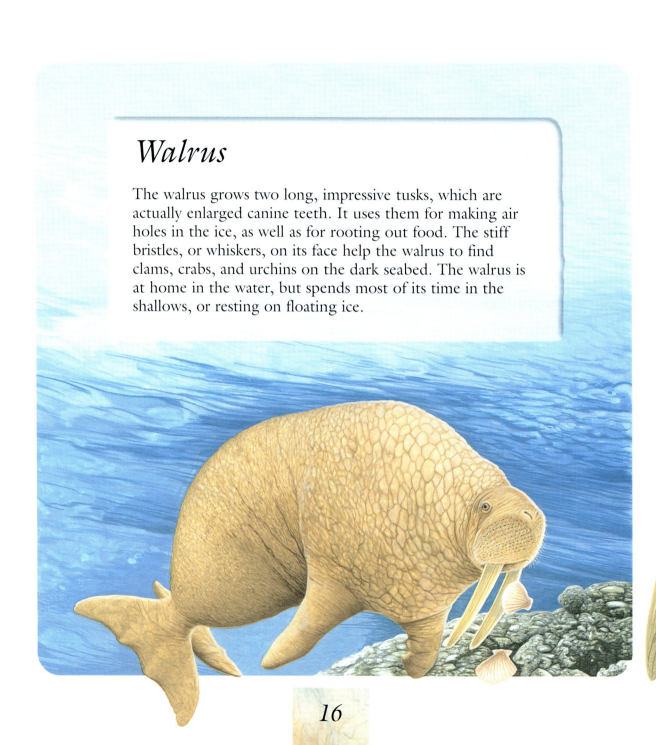

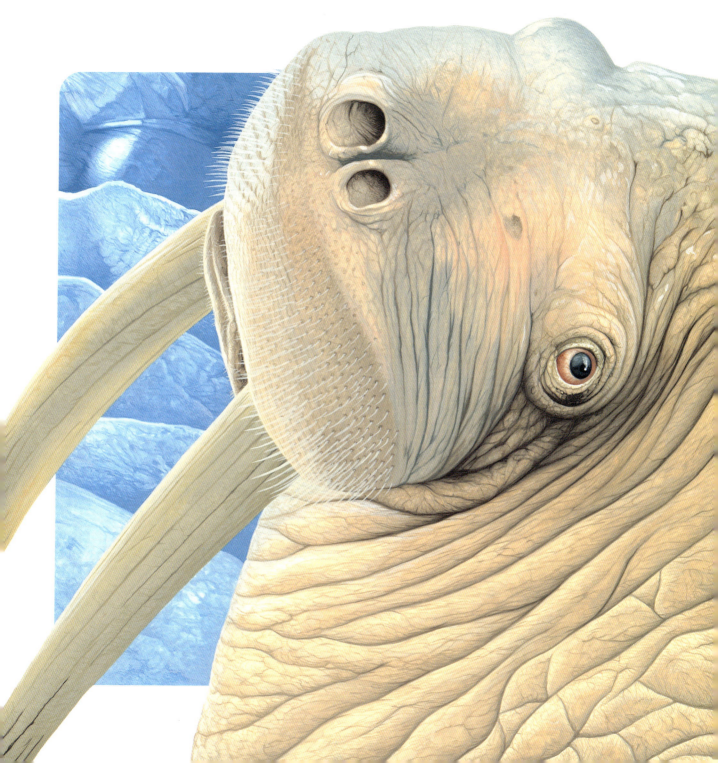

Polar Bear

Any creature spending time on the ice has to watch out for the great hunter, the polar bear. This large animal lives mainly off of seal meat. The polar bear's excellent sense of smell can catch the scent of food several miles away. Even seals hiding in dens under many layers of snow and ice are not safe from this hunter.

Killer Whale

The polar bear may rule the Arctic land and the great slabs of floating ice, but the killer whale is king in the water. Also known as orcas, killer whales are large dolphins. They prey on squid, seals, and even other dolphins and whales. These fierce animals have jaws lined with sharp teeth. They use their strong, blunt snouts to break through the ice to catch the seals that make up their diet. Killer whales also live at the opposite end of the world, in the waters of the Antarctic.

Ice Fish

Sharing the freezing Antarctic waters with the killer whale is the ice fish. This fish has a beak-like face, thick lips, and a large mouth. Unlike most other fish, it has no scales and is able to swallow large fish whole, which conserves energy. This is an advantage in the low Antarctic temperatures, especially during the winter. Lying on the seabed, the ice fish opens its large mouth to swallow passing fish such as the Antarctic cod.

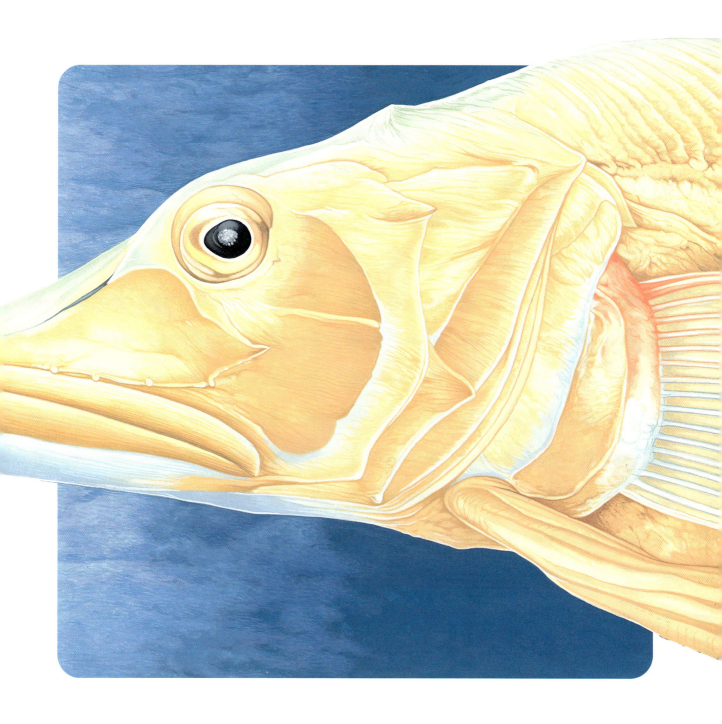

Gray-headed Albatross

Above the southern oceans flies the albatross. This large seabird can range in size from the huge wandering albatross, with a wingspan of up to 11 feet, to the smaller gray-headed albatross, with a wingspan of up to seven feet. The gray-headed albatross has a powerful, curved beak, trimmed in bright yellow and pink colors. These birds use their bills to clean each other's feathers. They pair for life and return to the same nesting site every year.

Elephant Seal

Other animals drawn to the Antarctic during the warmer months include the elephant seal. The name comes from its long, gray nose, which looks a bit like an elephant's trunk. Bull elephant seals fill their trunks with air to make a loud noise that warns other males off their territory. Hardy as these huge seals are, even they cannot endure winter near the South Pole. They spend the coldest months farther north, in warmer waters.

Emperor Penguin

One of the few animals to stay on the Antarctic mainland through the bitter winter is the emperor penguin. In order to stay warm, the emperor penguin has thick layers of fat on its body and air bubbles under its scale-like feathers that keep heat from escaping. This bird breeds and rears its young even during the coldest months. The emperor penguin protects its eggs and chicks from the icy temperatures by holding them on its feet under a flap of feathers.

Facts Behind the Faces

The animals in this book live in the extreme polar regions of the planet—in a world of snow and ice. This part of the book tells you more about these animals—where they live, their close relations, their eating habits, and their biggest enemies. From the woolly-coated musk ox in the north, to the hardy emperor penguin in the south, here are the facts behind the faces.

Musk Ox
Family: The musk ox family is related to cattle, water buffalo, and bison.
Where they live: Northern Canada, Alaska, Greenland, and Norway.
What they eat: Plants, especially grass.
Enemies: Wolves and human beings; musk oxen nearly died out in the 60s.
Size: Males can weigh up to 880 pounds.

Reindeer
Family: The deer family includes red deer, elk, and moose, and is similar to the caribou of North America.
Where they live: Across northern Europe and Asia.
What they eat: Grasses and leaves in summer, lichens in winter.
Enemies: Wolves, wolverines, lynx, and human beings.

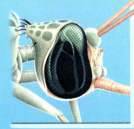

Mosquito
Family: The two-winged insect family includes flies and midges.
Other mosquitoes: There are over 3,000 different kinds of mosquitoes.
What they eat: Plant sap. Females sip blood to help their eggs grow.
Enemies: Human beings have developed chemicals to kill mosquitoes.
Special features: Female mosquitoes can carry diseases, such as malaria.

Snowy Owl
Family: The owl family includes screech owls and long-eared owls.
Where they live: They breed in the Arctic and move south in winter.
What they eat: Small mammals, including lemmings and hares.
Enemies: Human beings; owls are now protected by law in some places.
Special features: When threatened, young owls "play dead."

Arctic Fox
Family: The dog family includes foxes, wolves, and jackals.
Other foxes: Red foxes, fennecs, and bat-eared foxes.
Where they live: Coastal regions and islands of the Arctic Ocean.
What they eat: Small mammals, birds, and bird eggs.
Size: Around 20-24 inches in length, with a 12 inch tail.

Narwhal
Family: The white whale family includes narwhals and belugas.
Where they live: Coasts and rivers across the Arctic.
What they eat: Fish and squid.
Enemies: Human beings hunt the narwhal for its tusk.
Size: 11.5-16 feet long. Tusk can grow to 8.9 feet.

Walrus
Family: The walrus family includes the Pacific and Atlantic walrus and resembles the sea lion and fur seal.
Where they live: Arctic waters of North America, Europe, and Asia.
What they eat: Clams, crabs, and sea urchins.
Enemies: Polar bears and human beings.

Polar Bear
Family: The polar bear family is related to wolves and hyenas.
Other bears: Grizzly, black, spectacled, and sun bear.
Where they live: Arctic regions, on large stretches of floating ice.
What they eat: Seals, fish, birds, caribou, lemmings.
Size: 8-11 feet in length.

Killer Whale
Family: The killer whale is part of the dolphin family, despite its name.
Other dolphins: Bottle-nosed dolphin, white-sided dolphin, pilot whale.
Where they live: All oceans across the globe.
What they eat: Seals, penguins, dolphins, whales, and fish.
Enemies: Human beings keep them in captivity as performing animals.

Ice Fish
Family: There are 16 different types of ice fish in the Antarctic.
Where they live: In cold waters close to the South Pole.
What they eat: Other fish, which they can swallow whole.
Size: Up to 2 feet in length.
Special features: Blood is colorless, rather than red.

Gray-headed Albatross
Family: The albatross family includes petrels.
Other albatrosses: Black-footed, royal, sooty, and wandering albatross.
Where they live: Most oceans, except the North Atlantic.
What they eat: Fish, squid, and plankton.
Size: Wing-span of around 7.2 feet.

Elephant Seal
Family: The earless seal is related to Weddell and harbor seals.
Where they live: Around the Antarctic coast and islands.
What they eat: Fish and squid.
Size: Largest of the seal family, they grow up to 16 feet.
Special features: They can stay under water for as long as two hours.

Emperor Penguin
Family: Flightless bird family includes king, macaroni, and fairy penguins.
Where they live: The icy coast of the Antarctic.
What they eat: Fish, squid, and shellfish.
Size: The tallest penguin grows to around 39 inches in height.
Special features: They can endure winters in the Antarctic.

Index

Antarctic 20, 22, 26, 28, 32
Antlers 6
Arctic 6, 8, 10, 20, 31
Arctic fox 12, 31
Arctic Ocean 4, 12, 14, 31

Beak 24
Bear 18, 31
Bird 10, 24, 31, 32

Deer 6, 30
Dolphin 20, 32

Ears 12
Elephant seal 26, 32
Emperor penguin 28, 30, 32
Eyeballs 10
Eyes 10, 22

Fish 22, 31, 32
Fox 12, 31

Gray-headed albatross 24, 32

Horns 4, 6

Ice 4, 12, 16, 18, 20, 30, 31, 32
Ice fish 22, 32

Killer whale 20, 32

Lips 22

Mosquito 8, 30
Musk ox 4, 30

Narwhal 14, 31
Nose 26

Owl 10, 31

Penguin 28, 30, 32
Polar bear 18, 31

Reindeer 6, 30

Seal 18, 20, 26, 31, 32
Snow 10, 18, 30
Snowy owl 10, 31
Squid 20, 31, 32

Teeth 16, 18, 20
Tusk 14, 16, 31

Walrus 16, 31
Whale 14, 20, 31
Whiskers 16
Wolf 30, 31